McGRAW SCIENCE

Macmillan/McGraw-Hill Edition

Activity Workbook

GRADE 2

Mc Graw Hill **Macmillan McGraw-Hill**

New York   Farmington

Macmillan/McGraw-Hill

*A Division of The **McGraw·Hill** Companies*

Published by Macmillan/McGraw-Hill, of McGraw-Hill Education, a division of The McGraw-Hill Companies, Inc.,
Two Penn Plaza, New York, New York 10121. Copyright © by Macmillan/McGraw-Hill. All rights reserved.
No part of this publication may be reproduced or distributed in any form or by any means, or stored in a database
or retrieval system, without the prior written consent of The McGraw-Hill Companies, Inc., including, but not
limited to, network storage or transmission, or broadcast for distance learning.

Printed in the United States of America
Activity Workbook ISBN 0-02-280259-2/2
6 7 8 9 024 06 05 04

Table of Contents

Which of these are living?

What to do

1. **Classify** things in your classroom. List three or more living things. List three or more nonliving things.

2. Record the lists on a chart like this one.

_____

_____

Living	Nonliving

3. Share your chart with a partner. Tell how you knew how to classify each thing.

_____

Point it out!

In this activity, you will point out living and nonliving things.

What to do

1. Go outside with your class. As you walk around the school grounds, your teacher will point to different things.

2. Name things which are living. Name things which are nonliving.

3. Now, you will repeat the walk. This time, your teacher will call out the word *living* or *nonliving.*

4. Point to something that matches the word your teacher calls out.

What did you find out?

How did you know whether something was living or nonliving?

_____

Name _____

What do leaves need?

What to do

two potted plants

foil

1. Place both plants in a place that has a lot of light.

2. Cover the leaves of Plant B with foil.

3. Make a chart like this one. **Predict** what will happen to each plant. Write it in the chart.

_____

_____

	Plant A	Plant B
Prediction		
Day 1		
Day 2		
Day 3		
Day 4		
Day 5		

4. Record what you observe each day for a week. Keep plants moist.

5. Were your predictions correct? What do leaves need?

_____

Name _____

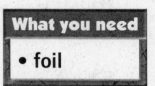

Warm it up!

In this activity, you will feel the difference between two kinds of sunlight.

What you need

• foil

What to do

1. Place one bare hand, palm up, in a sunny spot.

2. Cover the other palm with a piece of foil. Then put your foil-covered palm in a sunny spot.

3. Keep both palms in the sunlight for a few minutes.

What did you find out?

1. Did you feel any difference between your two palms?

_____

2. Which one felt warmer? Why?

_____

© Macmillan/McGraw-Hill

Which seeds grow into each plant?

What you need

picture cards 1–8

What to do

1. Sort the Picture Cards into two groups.
   Make one group for grown plants. Make another group for seeds.

Apple Seeds

2. Observe each group. **Infer** which seeds come from each plant. Match the cards.

3. Tell why you made each match.

_____

Name that seed!

In this activity, you will describe seeds from different kinds of fruits.

What to do

1. Your teacher will call out the names of different kinds of fruits.

2. Describe the seeds that each fruit protects. You can talk about how many seeds the fruit has and what size and color the seeds are.

What did you find out?

1. On another piece of paper, draw pictures of the different fruit seeds.

2. Look at the pictures you made. How are the seeds alike? How are they different?

_____

_____

What is made from plants?

What to do

1. Go on a plant hunt in your classroom.

2. **Infer** which objects are made from plants. Look for other objects that are not made from plants.

3. Fill in a chart like this one. Show which
group each item belongs to.

_____

_____

Made from Plants	Not Made from Plants

What am I?

In this activity, you will decide which things in your classroom come from plants.

What to do

1. Your teacher will give riddles about things in your classroom.

2. Answer each riddle. Then tell whether the thing in the riddle comes from plants.

What did you find out?

1. What part of a plant does each thing come from?

_____

2. Name some everyday things you use that come from plants.

_____

How can we classify animals?

zebra

rabbit

snake

fish

tiger

fox

What to do

1. Compare the animals in these pictures.

2. Make a diagram like this one. **Classify** the animals.

animals
with stripes

animals
with stripes
and fur

animals
with fur

3. Write the names of the animals in the correct parts of the circles. Tell how you decided where each animal belongs.

_____

_____

_____

4. Think of more animals that could belong to each group. Write them in your diagram.

_____

What's in a breed?

In this activity, you will discover how dog breeds are alike and different.

What to do

1. Look at the chart of different breeds of dogs. Notice how all the dogs are similar in some ways and different in others.

2. Name ways that all the dogs shown on the chart are alike.

3. Explain how all the dogs on the chart are different.

What did you find out?

Draw your own diagram showing some of the differences between the dogs on the chart.

_____

_____

What do pets need to live?

What to do

1. Think of a pet you have or would like to have. Draw a picture of it.

2. Imagine you had to leave your pet with a pet-sitter.

What you need

paper

crayons

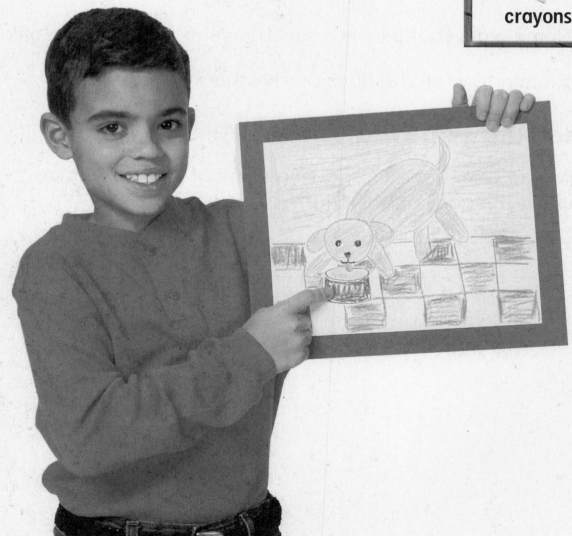

3. What does your pet-sitter need to do for your pet? Write sentences or draw pictures to **communicate** what your pet needs.

_____

_____

_____

What, where, how

In this activity, you will help create a chart about where an animal lives and how it gets food.

What to do

1. Together with your class, choose an animal. Your teacher will write this animal's name on the chart under the heading *What*.

2. Help describe where the animal that you chose lives. These facts will go on the chart under the heading *Where*.

3. Think about how this animal gets its food. Your teacher will write this information on the chart under the heading *How*.

What did you find out?

Choose a different animal. Fill in the chart below.

What	Where	How

How do animals grow and change?

What you need

picture cards
9–20

What to do

1. Group the Picture Cards for each animal.

2. Put the Picture Cards in **order** to show how each animal grows and changes.

3. Use the cards to tell about what happens first, second, third, and last.

_____

_____

_____

Stages of growth

In this activity, you will describe how animals change as they grow.

What to do

1. Look at the photograph of the adult animal.

2. Describe how the animal looked when it was first born.

3. Then discuss how the animal grew and changed as it became a young animal.

4. Tell about the changes that happened as this young animal grew into an adult.

What did you find out?

Compare the changes that this animal goes through to the changes a human goes through. How does a human grow and change from a baby into an adult?

_____

_____

Block Out!

What to do

1. Observe the two plants. Draw or write what you see.

2. Place both plants near a window. Tape all cracks on the box so that no light gets to the plant. Place the box over plant B.

3. How are the plants different after one week?

What you need

similar plants

cardboard box

masking tape

	Plant A	Plant B
Before activity		
After one week		

4. Why are they different?

_____

_____

Let It Grow

1. Plant each seed in a cup. Keep the soil moist. Put the cups in a sunny place.

2. Look at the plants every day.
   Draw changes you see in the plants.

3. What plant part did you see first?

   _____

   Second? _____

4. Describe the stems and leaves of the two plants.

   _____

   _____

What you need

2 seeds

soil

2 plastic cups

crayons

	Day 3	Day 5
Plant 1		
Plant 2		

Life Cycles

What to do

1. Put each animal's life cycle in order.
   Start the frog cycle with the egg card.
   Start the bird cycle with the newborn card.

What you need

picture cards
13-20

2. How are the two life cycles alike?

   _____

   _____

 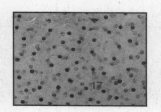  

_____ _____ _____ _____

3. How are the two life cycles different?

   _____

   _____

   _____

Name _____

Make a Pattern

When you see a pattern, you can predict what comes next. A pattern repeats things in a specific order.

What you need

crayons

drawing paper

What to do

1. Look at the pictures below. How many kinds of flowers are there?

   _____

2. Do you see the pattern? Use crayons to draw the pattern.

   _____

3. If you follow the pattern, what flower will be next?

   _____

4. Make a different pattern. Use at least 3 different objects.

   _____

Collect Data

You can collect data to find things you want to know. Data tells you about the world around you.

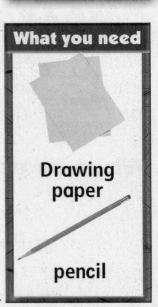

What you need

Drawing paper

pencil

What to do

1. Look around the classroom. What kinds of shirts are people wearing?

_____

2. Make a chart to record how many people are wearing cotton T-shirts.

T-Shirts	Other Shirts

3. How many people are wearing cotton shirts? How many are not?

_____

4. What other kinds of clothes are made from cotton?

_____

Put Events in Order

We put events in order to make sense out of things. We use special words like *first, next,* and *last.*

What to do

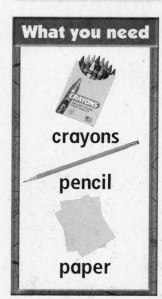
1. Look at the pictures below. Put them in order from first to last.

2. Draw pictures of 4 different things you did today. Put the pictures in order from first to last.

_____   _____   _____   _____

3. Write a story about your pictures. Attach extra pages if necessary.

_____

_____

4. Why do events in a story need to be in order?

_____

_____

Where do animals live?

What to do

What you need

- Picture cards 21-26

1. Sort the cards. Put the animals in one group. Put the places where the animals live in another group.

2. Match each animal with the place where it lives.

_____

_____

_____

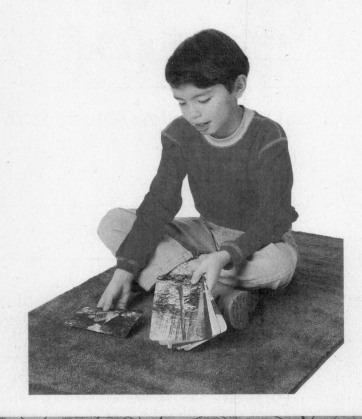

Explore
Activity
Lesson 1

3. Communicate why you matched the cards
the way you did.

_____

_____

_____

Where Animals Live

In this activity, you will discover places where animals live.

What to do

1. Look at the words on the chalkboard: *fish, nest, pond, rattlesnake, bear, desert, bird, cave.*

2. Which words are animals? Which words are places where animals live?

_____

3. Sort the animals and places where animals live into the chart below.

Animals	Places where animals live

What did you find out?

Which animals live where?

_____

_____

What is a forest like?

What to do

1. **Make** a **model** of a forest. Place a layer of soil and rocks in a large, empty bottle.

2. Place the plant in the soil. Water the soil and place the pill bug in the bottle. Wash your hands.

Materials

bottle

soil

plant

PLASTIC WRAP
plastic wrap

rocks

plastic spoon

pill bug

3. Cover the bottle with plastic wrap. Poke holes in it. Place it near a window.

4. What does your model tell you about a forest?

_____

_____

_____

Name _____

Woodland Forest Diagram

In this activity, you will make a diagram of a woodland forest.

What to do

1. Work in a group of seven. Each person will make one of the following: brown soil, gray rocks, black log, green plants, orange Sun, white cloud, blue drops of water.

2. Now put the pieces together to make a woodland forest diagram. Glue all of the parts in place on a piece of yellow paper.

What you need

- colored paper
- blunt safety scissors
- glue
- art materials

How can a rain forest animal find shelter?

What to do

1. Fold a paper plate in half. Glue six cotton balls inside the fold.

2. Stand up the plate like a tent. Put it in the tray. Pour a little water over the paper plate.

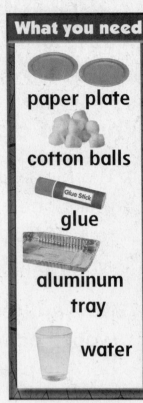

What you need

paper plate

cotton balls

glue

aluminum tray

water

3. What do you observe about the cotton balls? **Infer** why leaf bats find shelter inside rain forest leaves.

_____

_____

_____

Finding Shelter

In this activity, you will discover how to stay dry using shelters.

What to do

1. Sort all of the items into two groups. The first group of items would keep a person dry. The second group of items would not keep a person dry.

2. Compare and talk about your ideas. Then record your ideas by filling in the chart below.

What you need

- eraser
- newspaper
- pencil
- plastic bag
- umbrella

Dry	Not dry

What did you find out?

Think about what you have found out about shelter. Why do leaf bats make their homes under leaves?

How does the shape of a leaf help a plant?

What to do

What you need

paper towels

scissors

water

PLASTIC WRAP

plastic wrap

1. Cut two leaf shapes from the paper towels. Twist one leaf into the shape of a needle.

BE CAREFUL! Scissors are sharp!

2. Place both leaf shapes on plastic wrap. Wet them both.

3. Check both leaf shapes every 15 minutes.

4. Which leaf shape stayed wet longer? **Draw a conclusion** about which kind of leaf you might find in a dry place.

_____

_____

Name _____

Drying Leaves

In this activity, you will discover how fast different leaves dry out.

What to do

- leaves of different shapes

1. Put different types of leaves in a sunny spot. Guess which leaves will dry out the fastest.

2. Now observe and record the drying times of each leaf.

3. Show the results in the chart below.

Type of leaf	Drying time

How can color help animals hide?

What to do

1. Fold the white paper. Spread out the circles on the inside fold. Cover the circles.

2. **Predict** which color circles will be easier to see and pick up.

What you need

white paper

20 white circles

20 brown circles

3. Your partner will uncover the circles for ten seconds. One at a time, pick up as many circles as you can.

4. How many circles of each color did you pick up? Record these numbers. How does color make it easier or harder to pick up the circles?

_____

White circles	Brown circles
_____	_____

Name_____

Alternative Explore
Lesson 5

Circle Hunt

In this activity, you will discover how different colors help things hide.

What you need

- six colored circles

What to do

1. Stay in your seat as you look around the room for ten seconds. Try to find six colored circles.

2. Point to each circle as your teacher names its color.

What did you find out?

Which colored circles were easiest to find?

_____

_____

How does a duck stay dry?

What to do

1. Draw two ducks on paper. Cut them out.

BE CAREFUL!   Scissors are sharp!

2. Use a crayon to color both sides of one duck.
   Press hard. Make sure no white space is left
   on the paper. Do not color the second duck.

What you need

paper

scissors

crayons

cup of water

spoon

3. Drip water on one side of each paper duck.
Use a spoon. Which one stayed drier? Why?

_____

_____

4. Infer how a duck's feathers help it stay dry.

_____

_____

_____

How Some Water Birds Eat

In this activity, you will learn how ducks catch and eat food.

What to do

1. Place the small plastic beads in the bowl of water.

2. Imagine that the slotted spoon is a duck's beak. Many ducks eat by straining tiny bits of food out of water.

3. Try to catch the beads in the same way that a duck would catch food.

What lives in a salt water habitat?

What to do

1. Fill each container with two cups of water. Add two teaspoons of salt to one container. Mix it.

Materials

brine shrimp eggs

2 clear containers

spoon

salt

measuring spoon

measuring cup

hand lens

Fresh   Salt

©Macmillan/McGraw-Hill

2. Add 1/4 teaspoon of eggs to the fresh water.
Add the same amount to the salt water.

3. Check each container every day. **Observe** what happens.
Use a hand lens.

_____

4. Can brine shrimp grow in both containers? Tell why
or why not.

_____

_____

Food Chain Song

In this activity, you will sing a song to learn about an ocean food chain.

What to do

To the tune of "Farmer in the Dell," sing this song:

The fish are in the sea,
The fish are in the sea,
Hi-ho the dairio,
The fish are in the sea.

A shrimp eats small plants,
A shrimp eats small plants,
Hi-ho the dairio,
A shrimp eats small plants.

A small fish eats the shrimp,
A small fish eats the shrimp,
Hi-ho the dairio,
A small fish eats the shrimp.

A squid eats the fish,
A squid eats the fish,
Hi-ho the dairio,
A squid eats the fish.

A bigger fish eats the squid,
A bigger fish eats the squid,
Hi-ho the dairio,
A bigger fish eats the squid.

A seal eats the fish,
A seal eats the fish,
Hi-ho the dairio,
A seal eats the fish.

A whale eats the seal,
A whale eats the seal,
Hi-ho the dairio,
A whale eats the seal.

The whale is at the top,
The whale is at the top,
Hi-ho the dairio,
The whale is at the top.

What can oil do to a bird's feathers?

What to do

1. Put the feather in the water. **Predict** what will happen if you pour oil into the water.

2. Pour oil into the water.

What you need

tray of water

feather

oil

cup of oil

paper
towels

3. Tell what happens to the feather.

_____

4. Try to remove the oil from the feather and the water. Use a paper towel. Is it easy or difficult to do? Wash your hands when done.

_____

_____

Name _____

A Few Drops Go a Long Way

In this activity, you will discover how small amounts of oil can pollute big amounts of water.

What you need

- pie pan
- water
- oil
- hand lens
- flashlight

What to do

I. Describe the water in the pie pan.

_____

2. Watch as your teacher pours a few drops of oil on the water. Describe what you see.

_____

What did you find out?

How can a small amount of oil pollute a lot of water?

_____

_____

Name _____

Science Center

Card 7

Bug Box

What to do

1. Fill the carton halfway with soil. Add grass and a pill bug.

2. Put plastic wrap over the top of the carton. Use a rubber band to hold it.

3. Observe the pill bug. What does it do? Record what you see.

Materials

- milk carton

soil

pill bug

- grass

PLASTIC WRAP

plastic wrap

rubber band

pencil

What I see

Name _____

How Much Water?

What to do

What you need

soils

cups   bowls

hand lens

water

measuring cup

goggles

1. Use the hand lens to observe the soils. Describe how they look and feel.

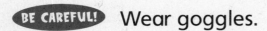

_____

_____

2. Which soil would you find in a forest? In a desert?

_____

3. Place the soils into the cups. Hold each cup over a bowl. Pour water into the cups.

   BE CAREFUL! Wear goggles.

4. Which bowl has more water in it? Which soil holds more water?

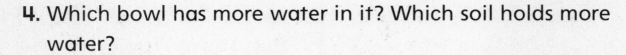

_____

Follow Directions

When you follow directions, things happen the way they are supposed to.

What to do

What you need

aluminum trays

soil

plant

water

cups

goggles

1. Put soil on one side of Tray 1. Put a plant with soil on one side of Tray 2.

2. **BE CAREFUL!** Wear goggles. Water the soil in Tray 1. Water the soil in Tray 2 the same amount.

3. Record what happens to the soil in each tray. Why do you think this happened?

_____

_____

Soil in Tray 1	Soil in Tray 2

Put Events in Order

When you put events in order, they make sense.
One card follows another card.

**Picture Cards
27–34**

What to do

1. Look at the season cards. Put the seasons in order starting with winter.

2. Look at the snowshoe hare cards. Match the hare cards to each season. Hint: A snowshoe hare's fur changes color on its feet first.

3. What color is the snowshoe hare's fur in winter?

_____

in summer?

_____

4. How did you choose the card of the snowshoe hare for spring?

_____

_____

_____

Paper Pond

What to do

1. List parts of a pond.

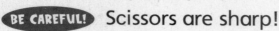

_____

_____

2. Decide what you need to make a paper pond.

3. Cut out the pond parts.

   BE CAREFUL! Scissors are sharp!

   Glue each part down.

4. Which pond parts are living? Which are nonliving?

What you need

paper

crayons

glue

scissors

Living things	Nonliving things

Make a Definition

A definition tells you what something is. Can you make a definition for salt water?

What to do

What you need

- salt water
- cotton swab

1. Use your senses to describe the salt water.

How It Looks	How It Feels	How It Sounds	How it Tastes	How It Smells

2. Write a new definition based on what your senses tell you.

_____

_____

3. What is your favorite fruit? Write a definition of the fruit based on what your senses tell you.

_____

Where does water for rain come from?

What to do

What you need

goggles

cup

sand

cup of water

plastic bag

1. **BE CAREFUL!** Wear goggles. Put sand into an empty cup. Add water until the sand is damp.

2. Put the cup into a plastic bag. Seal it. Put the bag in a sunny place.

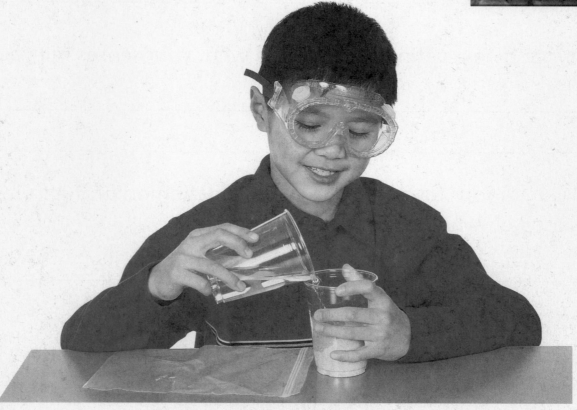

3. After a few hours, observe the bag.
What do you see?

_____

4. Draw a **conclusion** about what happened.

_____

_____

Foggy Mirrors

In this activity, you will discover how warm air changes into fog when it hits a cool surface.

What to do

I. Breathe hard onto the hand mirror.
   What happens?

   _____

2. Where does the fog on the mirror come from?

   _____

   _____

3. What might happen in nature when warm air meets cool air?

   _____

How can you change rocks?

What to do

1. Look at the rocks with a hand lens. Rub them on sandpaper. **Communicate** what happened to the rocks.

_____

2. Put the rocks inside the jar of water. Close the lid tightly. Shake the jar for a few minutes.

What you need

rocks

sandpaper

plastic jar of water

hand lens

3. Look at the rocks with a hand lens.
Communicate what happened.

_____

Rock Collecting

In this activity, you will discover how some rocks can change.

What to do

1. Go on a rock-collecting walk with your class. Find two or three different kinds of rocks.

2. Rub each rock against the sidewalk. What happens to the rocks?

_____

_____

3. Why did different rocks give different results?

_____

How can Earth's surface change?

What to do

What you need

foil

foil pan cut in half

bowl

water

soil

spoon

1. Put the two halves of the pan together. Line it with foil.

2. Make mud out of soil and water. Spread it in the pan. Wash your hands.

3. Let the mud dry in a warm place until it is hard.

4. Move the two sides of the pan quickly against each other.

5. Observe what happens. How did the surface change?

_____

_____

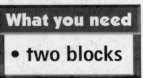

Earth Moves

In this activity, you will explore how Earth's surface can move.

What you need

• two blocks

What to do

1. Slide two blocks past each other in opposite directions.

2. Imagine the blocks are Earth's crust. What would happen if a road passed above the spot where the blocks moved apart?

_____

3. Use the blocks to explore other ways that Earth's crust could move during an earthquake. List all the different ways.

_____

_____

How can we get clues from prints?

What you need

clay

small objects

What to do

1. Flatten a piece of clay. Press a secret object into it. Gently take the object away.

2. Make prints with two more objects.

3. Trade clay prints with a partner. **Infer** what
objects made the shapes. What clues did you use to
figure them out?

_____

_____

Quick Prints

In this activity, you will discover how scientists find fossils in rock.

What to do

1. Describe what is left behind after your teacher presses an object in the clay.

   _____

   _____

2. Now look at the lump of clay covering a different object. Are you able to tell what is inside the clay?

   _____

3. How can scientists find fossils if they are hidden inside rocks?

   _____

   _____

Which bones fit together?

What to do

1. Cut out the bones.

   BE CAREFUL! Scissors are sharp!

2. **Make a model** of a dinosaur.
   Fit the bones together.

What you need

scissors

tape

large sheet
of paper

dinosaur
bone cutouts

3. Tape the bones together. Then tape them to a sheet of paper.

4. Tell how you put the pieces together.

_____

Dinosaur Dig

What to do

1. Divide into groups. One group waits while each other group takes a turn digging in the sand, trying to unearth one buried dinosaur bone.

2. Then, the last group fits all the bones together. Which dinosaur is modeled?

_____

3. What can the complete skeleton help tell you about the dinosaur?

_____

_____

What you need

- one set of *Tyranno-saurus rex* bones cutouts

- large box

- sand

© Macmillan / McGraw - Hill

What happens when animals can not meet their needs?

What you need

index cards
labeled
Food, Water
or Shelter

What to do

1. Line up with the class. Take three cards from the top of the pile.

2. If the three cards say Food, Water, and Shelter, go to the back of the line. If you are missing one of these cards, sit down.

3. Play for four more rounds.

4. What happens to the number of players each round? **Infer** what happens to animals when they can not meet their needs.

_____

Picture their Needs

In this activity, you will draw what an animal needs in order to live.

What to do

1. Fold the sheet of paper into four boxes.

2. Choose an animal. Then draw it in the upper left-hand box of the paper.

3. Think about three things that your animal needs to live. Draw these things in the other three boxes. Label your page "What a(n) _____ needs to live."

4. Compare and contrast your drawings with others. What are some things that all living things need to live?

_____

All Kinds of Rocks

What to do

1. Observe the rocks with a hand lens. How are they the same? How are they different?

_____

_____

What you need

hand lens

balance

rocks

crayons

2. Put one rock in each balance pan. Sort the rocks from lightest to heaviest.

```
┌────────┐ ┌────────┐ ┌────────┐ ┌────────┐
│        │ │        │ │        │ │        │
│        │ │        │ │        │ │        │
│        │ │        │ │        │ │        │
└────────┘ └────────┘ └────────┘ └────────┘
```

3. Which rock is the heaviest? Which is the lightest?

_____

Plant Roots Break Up Rocks

What to do

What you need

small cups

2 seeds

1. Gently push two bean seeds into the plaster. Make sure the seeds are just below the plaster.

2. Place your cup near the window. What do you think will happen overnight?

_____

3. What happened to your model rock the next day? Explain how plant roots can change rocks.

_____

_____

_____

Make a Model

What to do

What you need

sponge

scissors

paint

newspaper

tracking
card

1. Make a model of a dinosaur foot. Trace the footprint on sponge. Cut it out.

   BE CAREFUL! Scissors are sharp!

2. Choose a tracking card. It will tell you what dinosaur tracks to show.

3. Lightly dip your model in the paint. Wash your hands! Press it on the newspaper to make tracks.

4. Trade tracks with another group. What can you tell about the tracks?

Collect Data

When you track changes in weather, you collect data.

crayons

paper

What to do

1. Observe the weather in the morning, at lunch, and in the afternoon for 5 days.

2. Record what you observe on your weather chart.

	Monday	Tuesday	Wednesday	Thursday	Friday
Morning					
Lunch					
Afternoon					

3. How did the weather change during the day? How did it change during the week?

_____

4. Do you see any weather patterns? What are they?

_____

Use a Chart

When you put information in a chart, it is easy to compare two or more things.

What to do

I. Read about the two dinosaurs below.

Maiasaura means good mother reptile. It took care of its babies. It was 30 feet long and about 7 feet tall. Thousands of them herded together. Maiasaurs only ate plants.

Oviraptor means egg robber. Its strong jaw and toothless beak means it ate both plants and meat. It was about 6 feet long and 3 feet high. Oviraptors were also herders.

2. Put the dinosaur information on the chart.

	Traveled in herds	Size	Cared for young	What they ate
Maiasaura				
Oviraptor				

3. What is the same about the dinosaurs? What is different?

_____

_____

Name _____

Follow the Drop

What to do

1. Cut out the picture of the water cycle from your workbook. **BE CAREFUL!** Scissors are sharp! Glue it onto the plate. Cover the plate with wax paper.

2. Place one drop of water on the wax paper. Pull the drop around the water cycle with the straw until your teacher says, "Stop."

3. Which part of the water cycle is the drop in? Where in the water cycle is water a gas? Where is it a liquid?

_____

_____

What you need

paper plate

scissors

glue

wax paper

straw

crayons

How does day change to night?

What to do

1. **Make a model** of Earth. Push the pencil through the ball. **BE CAREFUL!**

2. Press the paper clip into one side of the ball.

What you need

- **foam ball**
- **unsharp-ened pencil**
- **paper clip**
- **flashlight**

3. Have a partner shine the flashlight at the paper clip.
Where is it day on your model?

_____

4. Slowly spin the ball with the pencil. What happens to the
paper clip? Tell how this model shows how day changes
to night.

_____

_____

_____

_____

Day or Night

In this activity, you will tell when a label is in daylight and when it is in darkness.

What you need

- basketball
- flashlight

What to do

I. Watch as your teacher spins the basketball. What happens when the ball is spun a half turn around?

_____

2. Say *day* when the label is in daylight. Say *night* when the label is in darkness.

3. Explain how this shows day changing into night.

_____

_____

_____

Name _____

How does Earth move through the year?

What to do

What you need

scissors

crayons

Sun
worksheet

1. Color the Sun yellow on your worksheet. Color Earth blue.

2. Cut out the picture of Earth along the dotted lines.

BE CAREFUL! Scissors are sharp!

3. Move Earth along the arrows on the page.

4. **Compare** where Earth is during different months of the year. How does Earth move during the year?

_____

_____

_____

Earth's Orbit

In this activity, you will discover when Earth is closest to and farthest from the Sun.

What you need

• chalk
• large and small ball

What to do

1. On the classroom floor, trace the orbit that Earth travels around the Sun.

2. Point out the months when Earth is closest to the Sun. Then point out the months when Earth is farthest from the Sun.

_____

_____

3. On one half of a piece of paper, draw Earth orbiting closest to the Sun. On the other half, draw Earth orbiting farthest from the Sun.

What did you find out?

What shape path does Earth travel around the Sun? How much time does this journey take?

_____

_____

What makes moonlight?

What to do

1. Place a flashlight on a table. This is the Sun. Shine it at the large ball. This is Earth. Wrap the small ball in foil. This is the Moon.

2. Make the classroom dark except for the flashlight.

What you need

- large foam ball

- small foam ball

- foil

- flashlight

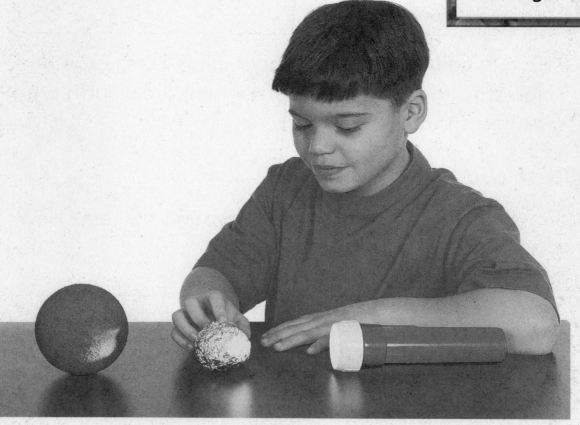

3. Move the Moon in a circle around Earth. Observe what makes the Moon shine. **Infer** what makes moonlight.

_____

By the Light of the Moon

In this activity, you will discover how light from the Moon reaches Earth.

What to do

1. Take turns standing where you can see the light of the mirror Moon.

2. Where does this moonlight come from?

_____

3. How does light from the Moon reach Earth?

_____

Vocabulary

- round hand mirror
- flashlight
- stack of books

How does the Moon change over time?

What to do

1. Look at the Picture Cards of the Moon.

2. Put the pictures **in order**. Start with the card that shows the least amount of the Moon.

- Picture Cards 35–38

3. Tell what the Moon looks like from the first
card to the last.

_____

_____

4. What do the cards tell you about how the Moon changes
over time?

_____

Name _____

Moon model

In this activity, you will discover shapes that the Moon makes when it is lit by the Sun.

What you need

- large ball
- flashlight

What to do

1. Watch as the flashlight Sun shines on the Moon ball.

2. Draw the shape that you see. What kind of Moon is this shape called?

_____

3. Now draw the other shapes that the lit part of the Moon makes.

4. Why does the Moon seem to change over time?

_____

What does the night sky look like?

What to do

What you need

- cereal box
- black paper
- tape
- flashlight
- scissors

1. Wrap a box in black paper. Carefully poke holes on one side of the box to show part of the night sky.

   BE CAREFUL! Handle scissors carefully.

2. Cut a hole in one end of the box. Shut off the lights. Shine a flashlight into the box.

Name_____

3. Observe your box and share with others.
   How does the box look like the night sky?

_____

_____

Where are stars during the day?

In this activity, you will discover why stars aren't visible during the day.

What to do

1. Describe the difference in the picture from the projector, with the lights on and off.

_____

2. Which do you think is brighter—light from the stars or sunlight?

_____

3. How can the picture from the projector with the lights on be compared to the sunlight?

_____

_____

What did you find out?

When are the stars in the sky?

_____

How are orbits alike and different?

What to do

1. Work in a group. Put a chair labeled **Sun**, in the center of an empty room.

2. Tape a straight line from the chair to a wall. Place numbers 1 to 9 in order along the tape. Have each person line up on a number.

3. **Make a model** of an orbit. Each person walks in a circle around the chair. Take the same size steps. Count your steps together.

What you need

- chair
- masking tape
- string

4. Who gets back to the tape first? How were the orbits alike and different?

_____

_____

_____

_____

_____

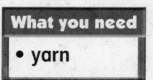

Yarn Orbit Model

In this activity, you will make a model of planets orbiting the Sun.

What you need
• yarn

What to do

1. Cut the yarn into three different lengths. Tie a knot in each length of yarn to make orbiting circles.

2. Now compare each of the three lengths of yarn. Which one would take the shortest amount of time to travel around the Sun? Which would take the longest amount of time?

   _____

   _____

3. How are these orbiting circles alike? How are they different?

   _____

Sun Shadows

What to do

What you need

pencil

ball of clay

chalk

ruler

paper

1. Stick the pencil point in the clay.

2. Place the clay on a piece of paper. Put it outside where you can see the pencil's shadow on the paper.

3. Trace the pencil's shadow on the paper three different times during the day.

4. Compare the three shadows. Why did the shadows change?

_____

_____

_____

_____

Read a Bar Graph

You get information when you read a bar graph.

What to do

1. Read the bar graph. What information does it give you?

   _____

2. How many months are on the bar graph?

   _____

3. How many nights in February were clear?

   _____

4. Which two months had the same number of clear nights?

   _____

5. Which month had the greatest number of cloudy nights?

   _____

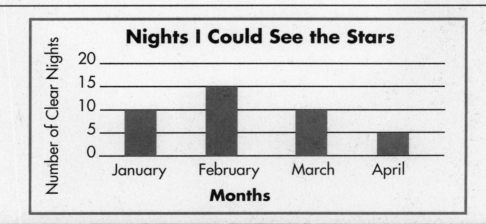

Nights I Could See the Stars

Number of Clear Nights

20
15
10
5
0

January   February   March   April

Months

Star Shine

What to do

What you need

flashlight

1. Shine the flashlight onto a wall in a dark room. What happens?

_____

2. Go outside with an adult. Shine the flashlight onto a wall. What happens?

_____

3. Why do you think you can't see all the stars during the day?

_____

Predict Events

When you predict events, you make a guess about what will happen next. You use what you know to help you predict.

What to do

1. At night, stand with an adult in a spot where you can see the Moon. Draw what you see. Use one marker to draw the Moon.

2. After one hour, check the Moon from the same spot. Draw where the Moon is.

3. Predict where the Moon will be in one more hour. Draw your prediction with the other marker.

4. Check to see if your prediction was correct.

Moon Watch

What to do

What you need

index cards

jar lid

1. Use the jar lid to draw a Moon on your Moon Lookout Card.

2. Your teacher will tell you what date to write on the card.

3. On your night, look at the Moon with an adult. Fill in the shape of the Moon that you see onto your Moon Lookout Card.

4. Tape your Moon Lookout Card on the calendar. How does the moon change during the month?

_____

_____

_____

Use a Chart

You use a chart to get information. Information is easy to read when it is in a chart.

What you need

• Weather section of newspaper

What to do

1. Look at the two charts. What do they tell you? _____

Summer Days

	Monday	Tuesday	Wednesday	Thursday	Friday
Sun Rises	5:03 A.M.	5:04 A.M.	5:04 A.M.	5:05 A.M.	5:05 A.M.
Sun Sets	7:05 P.M.	7:05 P.M.	7:05 P.M.	7:05 P.M.	7:05 P.M.

Winter Days

	Monday	Tuesday	Wednesday	Thursday	Friday
Sun Rises	6:41 A.M.	6:41 A.M.	6:42 A.M.	6:43 A.M.	6:44 A.M.
Sun Sets	5:00 P.M.	5:00 P.M.	5:00 P.M.	5:00 P.M.	5:00 P.M.

2. Why are the days longer in the summer than in the winter?

_____

3. How many hours of sunlight do you have right now? Look up the sunrise and sunset times all this week in the newspaper. _____

4. Make a chart to show the information you find.

	Monday	Tuesday	Wednesday	Thursday	Friday
Sun Rises					
Sun Sets					

How can you tell what is inside?

What to do

What you need

6 containers

1. **Observe** each container without opening it.

2. Guess what is inside each one. Record your guess.

Explore
Activity
Lesson 1

3. Open each container to see what is inside.
What clues did you use to tell what was inside?

_____

_____

_____

_____

What's around you?

In this activity, you will use different senses to name what is around you.

What to do

1. Work in pairs. One partner will close his or her eyes. Then he or she will listen carefully and describe everything that can be heard.

2. Now switch roles. The other partner will close his or her eyes and cover both ears with the hands. Using his or her nose, the partner will describe everything that can be smelled.

3. How did your sense of hearing or smell help you discover what's around you?

_____

_____

_____

How can you put matter in order?

What to do

What you need

classroom objects

balance

1. Look at each object. Which has the most mass? Which has the least mass? Predict.

_____

2. Compare two objects on the balance.
The one that makes the pan go lower has more mass.

3. Put the objects **in order** from least mass to most mass.

_____

_____

Name _____

Alternative Explore

Lesson 2

Least and Most

In this activity, you will guess and compare masses of objects.

Vocabulary

- small classroom objects
- balance

What to do

1. Look at the group of objects. Guess which has the least mass. Then guess which has the most mass.

Least mass:

_____

Most mass:

_____

2. Now use the balance to compare the masses of two objects at a time. Record your results.

3. Put all of the objects in a line, from smallest mass to greatest mass. Were your guesses correct?

_____

How can you change matter?

What to do

1. Observe your objects. Think of ways to change them and put them together.

2. **Investigate** how to change and put together the objects. Make a plan and try it out. **BE CAREFUL!** Scissors are sharp!

_____

_____

_____

What you need

glue

paper

scissors

craft materials

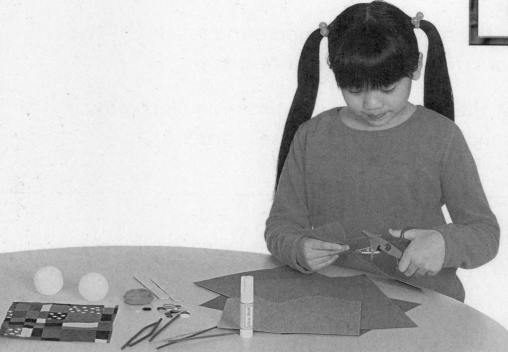

Name _____

Explore
Activity
Lesson 3

3. What did you make? What did you do to make it?
Tell about the different ways you changed matter.

_____

What's it made of?

In this activity, you will guess all of the different materials used to create a sculpture.

What to do

1. Examine the sculpture that your teacher made. Use your senses of sight and touch to inspect the sculpture.

2. List all of the different materials that were used to create the sculpture.

3. What kinds of matter were used in the sculpture: solid, liquid, or gas?

_____

4. On a piece of paper, draw a sculpture using materials found in and on your desk.

5. Switch pictures of your sculpture with a partner. See if you can identify all the materials in each other's sculpture.

Name _____

How can heat change matter?

What to do

1. Find a sunny spot outside on a warm day. Place the ice cube, the butter, and the chocolate on top of the paper plates. Draw how they look.

What you need

paper plates

ice cube

butter

chocolate

2. Do you think the Sun will change any of the items?
Why or why not?

_____

3. Leave the paper squares in the Sun for 20 minutes.

4. **Communicate** what happens to each item.
Draw how they look. How did heat change matter?

_____

_____

_____

Name _____

Alternative
Explore
Lesson 4

Effect of Heat

In this activity, you will feel how heat changes matter.

What you need

• **desk lamp (if needed)**

What to do

1. Place one of your hands in the sunshine or under the desk lamp if it's a cloudy day.

2. How does your hand in the sunlight feel, compared to your other hand?

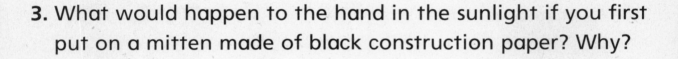

_____

3. What would happen to the hand in the sunlight if you first put on a mitten made of black construction paper? Why?

_____

How does light move?

What to do

1. Work with a partner. Stand near a wall with the flashlight.

2. Your partner will stand a few feet away from you and hold the mirror.

What you need

flashlight

mirror

3. Turn on the flashlight and shine it at the mirror. Your partner will try to make the light shine from the mirror onto the wall.

4. Observe what happens to the light. What did you find out about how light moves?

_____

_____

_____

Name _____

Bouncing Light

In this activity, you will explore how sunlight bounces.

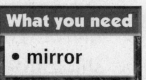
What you need
- mirror

What to do

1. Watch how your teacher uses a mirror to reflect the sunlight.

2. Describe what is happening to the sunlight.

_____

3. In what shape does the sunlight travel from the mirror to the building?

_____

4. What would happen on a cloudy day?

_____

How is sound made?

What to do

paper cup

string

goggles

paper clip

1. Work with two partners. Make a tiny hole in the bottom of the cup.

2. Tie the string to the paper clip. Pull the string through the hole until the clip is tight inside the bottom of the cup.

3. You and a partner hold the cup and string. A third partner snaps the string. **BE CAREFUL!** Wear goggles.

4. Observe what happens. How did you make sound?

_____

Making Vibrations

In this activity, you will explore how sounds vibrate through the air.

What to do

1. Watch and listen as the stretched rubber band is plucked.

2. Describe what you see and hear.

   _____

3. Now watch and listen as the experiment is repeated with a different sized rubber band. Describe how the sound changes.

   _____

4. What happens to the sound when the rubber band is stretched out farther?

   _____

Predict

When you predict, you make a guess. Your guess is based on what you think will happen.

What to do

What you need

cold water

warm water

food coloring

safety goggles

1. **Predict** what will happen to food coloring in a pan of warm water. What will happen in cold water?

   _____

2. Put three drops of food coloring in each pan of water. Draw what happens.

   BE CAREFUL! Wear goggles.

	Cool Water	Warm Water
What the food coloring does in water		

3. Was your prediction correct?

   _____

4. What did you observe about food coloring in warm and cold water?

   _____

Make a Bubble

What to do

1. Dip the bottle upside down in the soap. Be sure a bubble goes across the opening.

2. Put the bottle in the warm water bottom side first. What happens?

_____

3. Put the bottle in the cold water. What happens?

_____

4. What makes the bubble get bigger? What makes it get smaller?

_____

_____

_____

What you need

plastic bottle

hand soap

warm water

cold water

Make a Prediction

When you make a prediction, you use what you know to decide what will happen.

What to do

white construction paper

black construction paper

flashlight

1. Will a light shine brighter on white or black paper? Make a prediction.

_____

2. Make the room dark. Shine the light on the white paper. Shine it on the black paper.

3. Where was the light brighter?

_____

_____

Why do you think so?

_____

_____

Ask a Question

When you ask a question, you try to learn more about something.

What to do

What you need

4 plastic bottles

water

1. Fill the bottles with different amounts of water.

2. Blow across the top of each bottle. What kinds of sounds do you hear?

	Bottle with a little bit of water	Bottle just under half-filled	Bottle just over half-filled	Bottle almost full
Sound I hear				

3. Ask two questions about sound and the bottles of water.

_____

_____

_____

_____

A Rubber Band Guitar

What to do

What you need

1. Cut out a circle on the shoebox lid.
   BE CAREFUL! Scissors are sharp! Wrap four different rubber bands around the shoebox.

rubber bands

2. Pluck each rubber band.
   BE CAREFUL! Wear goggles.

   What do you hear?

   _____

   _____

shoebox

scissors

3. Line up the rubber bands in order from highest sound to lowest sound.

safety goggles

4. Which rubber bands make high sounds?

   _____

   Which make low sounds?

   _____

A Sound Game

What to do

I. Work with a partner. Choose an animal sound to make.

2. Stand in different parts of the room. Close your eyes. Your partner will make the animal sound.

3. Have someone walk with you. Walk until you find your partner. How did you find your partner? What animal sounds did your partner make?

_____

4. Think about different animals and the sounds they make. Fill in the chart.

Animal	The Sound It Makes
Cat	meow
Dog	
Cow	
Frog	
Snake	
Crow	

How far can different things move?

What to do

1. Mark a starting line with tape. Line up the objects.

2. Tap each object to make it move forward. Do not hit one thing harder than the others.

Name _____

Explore Activity
Lesson 1

3. Measure how far each object moved. Use a ruler.

4. Record how far each object moved. Which object moved the farthest? Why do you think so?

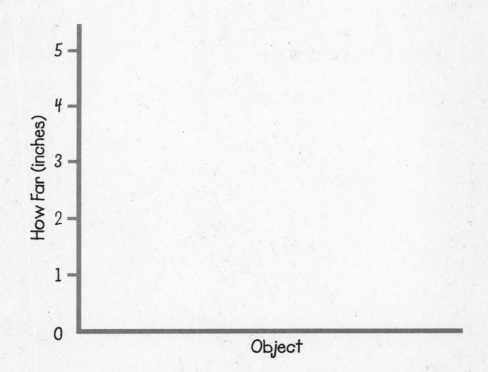

Name _____

Mother, May I?

In this activity, you will play a game following directions based on measurements.

What you need
• inch rulers
• yardstick

What to do

1. Play the game "Mother, May I?" Follow the directions using measurements that your teacher gives you.

2. Use an inch ruler or yardstick to measure all your movements.

What did you find out?

Which movement was the shortest? Which was the longest?

_____

How can you slow down a force?

What to do

1. Stack the books. Put the edge of the board on the books to make a hill.

What you need

cardboard

3 books

toy car

sandpaper

cloth

tape

2. Put the car at the top of the hill and let go.
Do not push it. Place tape where the car stopped.

3. Cover the board with wax paper. Repeat step 2.

4. Try the activity again with sandpaper, then cloth.

5. **Compare** how far the car went each time. What slowed down the car most?

_____

Rough and Smooth Sliding

In this activity, you will test objects to see which ones slide best.

What you need
• classroom objects

What to do

1. Test four objects to see which slide best.
   One by one, slide the objects across your desk.

2. Use the box below to compare the textures of the objects.

Object	Rough	Smooth
1.		
2.		
3.		
4.		

What did you find out?

Which objects slide best?

_____

Name _____

How can force help you lift things?

What to do **BE CAREFUL!** Wear goggles.

What you need

book

2 pencils

tape

goggles

1. Tape a pencil to your desk. Place the book next to the pencil.

2. Look at the pictures below. Which looks like the easiest way to lift the book?

_____

A

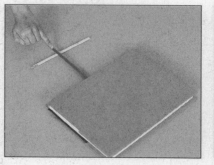

B

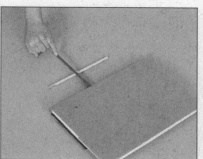

C

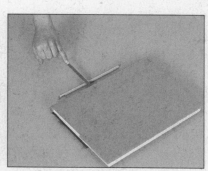

3. **Investigate** which way is easiest. Make a plan.
Then try it out.

4. What was the easiest way to lift the book? Why do you
think so?

_____

_____

Lifting Things

In this activity, you will make a lever to lift blocks.

What to do

1. Make your own lever. First, tape down the cylinder to keep it from rolling. Next, center the ruler across the cylinder.

2. Now put one block on the far end of the ruler. Use your hand to carefully press down on the near end of the ruler.

3. Add blocks, one at a time, to the lever. Record your results.

Number of Blocks	Blocks Lift?

What did you find out?

Compare results. What was the greatest number of blocks that could be lifted with the lever?

_____

How can you use less force?

What to do

1. Tie 10 washers onto the end of the string of the Puller Pal.

   BE CAREFUL! Wear goggles.

2. Use the Puller Pal to lift the washers straight up. **Measure** how far the rubber band stretches.

What you need

washers

Puller Pal

books

cardboard

goggles

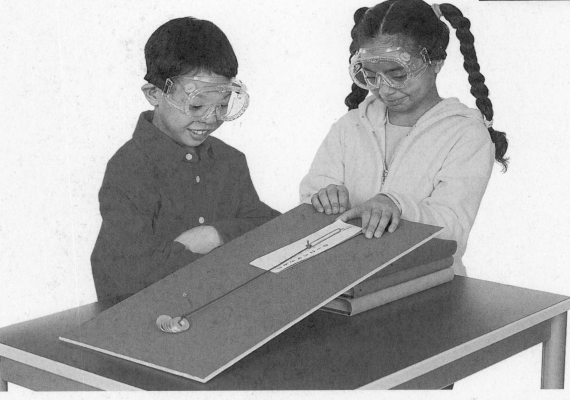

3. Place some books under one end of the board to make a hill. Pull the washers up the board. Measure how far the rubber band stretches.

4. When was less force needed to move the washers?

_____

Lift or Slide

In this activity, you will discover whether lifting or sliding objects is easier.

What to do

What you need

• books

• a small box of blocks or marbles

1. The first group of volunteers will take turns lifting the box. Place the box on top of the stack of books.

2. The second group of volunteers will take turns sliding the box up the ramp to the top of the stack.

3. Now switch groups. Repeat the lifting or sliding experiment.

What did you find out?

Which way was easier? Why?

_____

_____

What will stick to a magnet?

What to do

What you need

paper bag

small objects

magnet

string

pencil

1. Make a magnet fishing pole. Tie string to a pencil. Tie a magnet to the end of the string.

2. Put all the objects in a bag. Predict which objects will stick to the magnet.

_____

3. Use the fishing pole to fish out objects from the bag.

4. Classify each object to show whether it sticks to the magnet. List the objects on a chart.

_____

_____

_____

Sticks	Does Not Stick

To Stick or Not to Stick

In this activity, you will hunt for things
that will stick to the magnet.

What you need

• magnet

• 2 index cards

What to do

1. Make two tent labels with your index cards.
   Write *Will stick* on one card. Write *Will not
   stick* on the other card.

2. Hunt for things in your desk that you think will stick to
   the magnet.

3. List your guesses in the chart below.

4. Test your objects. Then sort them into either the *Will stick*
   or the *Will not stick* group. Use the tent labels to group
   the objects.

5. Check your chart. Draw a circle around your correct
   guesses. Mark an *X* through your incorrect guesses.

How can you make a magnet?

What you need

nail

bar magnet

paper clips

What to do

1. Stroke the iron nail in one direction.

BE CAREFUL! Nails are sharp!

2. Lift the nail at the end of each stroke before beginning another. Do this 50 times or more.

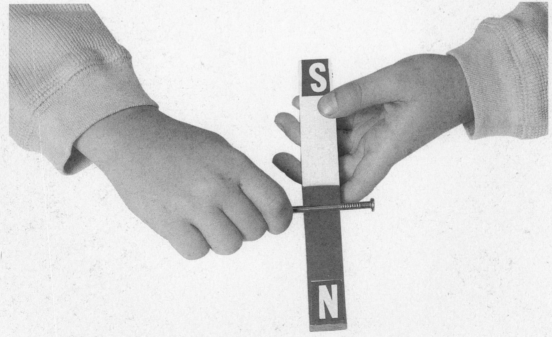

3. Test your nail. Can it pick up a paper clip?
Communicate what happens.

_____

4. What else can your magnet pick up?

_____

Name _____

Temporary Magnets

In this activity, you will make a magnet from a paper clip chain.

- **strong magnets**
- **paper clips**

What to do

1. Use the magnet to pick up a paper clip. Now use that paper clip to pick up another paper clip.

2. See how many paper clips you can pick up in a row.

What did you find out?

1. Explain what happens to the paper clip chain.

   _____

   _____

2. What happens when you separate the first paper clip from the magnet?

   _____

Which Way Do Roots Grow?

What to do

1. Fill the jar halfway with wet cotton balls.

2. Put the seeds between the jar and the cotton balls. Put the seeds in different positions. You should be able to see the seeds.

3. Label 4 pieces of tape with a number from 1 to 4. Place the tape on the outside of the jar under each seed.

4. Observe the roots of each seed. Draw how they grow. What happened to all the roots?

What you need

jar with lid

wet cotton balls

4 bean seeds

masking tape

pencil

	Seed 1	Seed 2	Seed 3	Seed 4
Day 1				
Day 2				
Day 3				
Day 4				
Day 5				

Name _____

Name _____

Science Center
Card 32

Make a Prediction

When you make a prediction, you make a guess. A prediction is based on what you know and what you think will happen.

What you need

cardboard ramp

3 books

toy truck

rocks

measuring stick

What to do

1. Make a ramp with cardboard and books.

2. Will the empty truck roll farther than the truck filled with rocks? Make a prediction.

_____

3. Roll the empty truck off the ramp. Measure how far it rolled.

_____

4. Repeat step 3 with rocks on the truck. Which truck rolled farther? Why? Was your prediction correct?

_____

_____

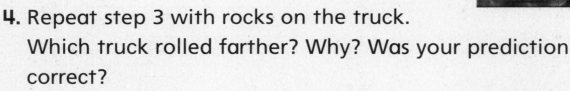

Make a Bar Graph

When you make a bar graph, you can compare things.

What to do

What you need

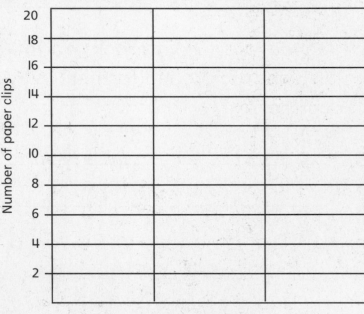

balance

3 small items

paper clips

1. Put the first item in the balance. Add paper clips to the other side until the balance is even. Record the number of paper clips on the chart. Repeat step I for the other two items separately.

2. Repeat step I for the other two items separatley.

3. Write the names of the three items on the bottom of the chart. Draw one bar for each item.

4. Which bar is highest? Which is lowest?

_____

_____

_____

Number of paper clips

20
18
16
14
12
10
8
6
4
2

_____ _____ _____

Item Name

Which Ramp?

What to do

1. Attach the milk carton to the puller pal.

   BE CAREFUL! Wear goggles.

2. Do you think it will take more force to pull an object up a short ramp or a long one?

   _____

3. Pull the carton up each ramp.
   Record the puller pal numbers.

   _____

What you need

short ramp

long ramp

heavy load

puller pal

3 books

goggles

	Short ramp	Long ramp
Number on the puller pal		

4. On which ramp did you use more force?

   _____

   How do you know?

Follow Directions

What to do

1. Hold 10 pages of your textbook together with a paper clip.

2. Lift the pages with the magnet.

3. Add more pages until the magnet can no longer lift them. How many pages did your magnet lift?

_____

_____

What you need

book

paper clip

bar magnet

Magnet Strength

What to do

1. Predict which magnet is the strongest.

   _____

2. Put the paper clip on the top line of the paper. Push one magnet slowly up the paper towards the clip.

3. Stop just as the clip begins to move toward the magnet. How many lines did the clip move?

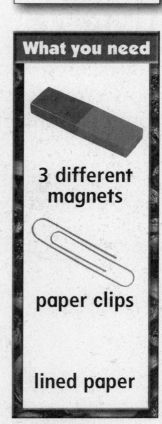

What you need

3 different magnets

paper clips

lined paper

Number of Lines			

4. Repeat steps 2 and 3 with the other magnets. Which magnet is the strongest? How can you tell?

   _____

   _____

Name _____

Observe

Your senses help you **observe** things. How? Your senses tell you how things look, sound, feel, smell, or taste.

What you need

pencil

crayons

paper

What to do

1. **Observe** something in the Science Center.

2. How does it look? Feel? Smell? Sound?

_____

_____

_____

3. Draw and write about it.

_____

4. What object did you observe? Which of your senses helped you the most?

_____

Measure

You can **measure** to find out how long, how fast, or how warm something is. You use numbers to record the answer.

2 cups of water

2 thermometers

What to do

1. Fill one cup with warm water. Fill the other cup with cold water.

2. Place a thermometer in each cup. Wait 2 minutes.

3. Record each temperature.

_____

4. What other things can you **measure** with a thermometer?

_____

Compare

You observe things to **compare** them. You find out how they are alike and different.

What to do

I. **Compare** the second-graders in the picture.

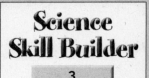

2. List three ways they are alike.

_____

_____

3. List three ways they are different.

_____

_____

Classify

You **classify** when you make groups that are alike in some way.

What to do

I. Look at the picture of the beans.

2. **Classify** the beans by size. Draw the two groups. Label them Big and Small.

3. Find another way to classify the beans. Write labels for the new groups.

Make a Model

You **make a model** when you do something to show a place or thing. A model helps you learn how a place looks or how a thing works.

What you need

paper

paper fastener

crayons

scissors

What to do

I. Make a **model** of a clock. Include numbers, and hands.

BE CAREFUL! Scissors are sharp!

2. What can you learn about the real thing
   from the model?

   _____

3. How is the real thing different from the model?

   _____

   _____

Communicate

You **communicate** to share your ideas. You can talk, write or draw to communicate.

What to do

I. Think about your favorite food.

2. Write a description of that food.

_____

3. **Communicate** to a friend. Read your description. Ask your
friend to name the food you described.

Infer

To **infer,** you use what you know to figure something out.

What you need

paper

pencil

What to do

1. Look at the pictures.

2. Use what you know to **infer** which place is warmer.

3. Write a short story to tell what games children can play in each place.

_____

_____

Put Things in Order

To put things **in order**, you tell what happens first, next, and last.

What to do

I. Think about all the things you did after you woke up this morning.

2. List each thing on your paper. Write them in **order.**

_____

3. Draw a picture of you doing one of those things.

Predict

You use what you already know to help you **predict** what will happen next.

"I'm hungry," said Laura.

"Me too," said Jack.

"I wish I had a snack," Laura said.

"All I have are these grapes," replied Jack.

What to do

I. Read the story above.

2. Predict what you think will happen next.

_____

3. Draw a picture to show it.

Investigate

To **investigate,** you make a plan and try it out.

What you need

pencils

clay

blocks

What to do

1. **Investigate** a plan for building a clay and pencil shape. It should be able to hold up at least 2 small blocks.

2. Try it out.

3. Place as many blocks as you can on your shape.
How many did it hold before it fell?

_____

Draw a Conclusion

To **draw a conclusion,** you use what you observe to explain what happens.

What you need

crayons

paper

What to do

I. Look at the picture. Where do you think the girl is going? What will she do there?

_____

_____

2. Draw a conclusion. Show it in a picture.

1

2

5

6

9

10

11

12

17

18

21

22

29

30

©Macmillan/McGraw-Hill

35

36

37

38

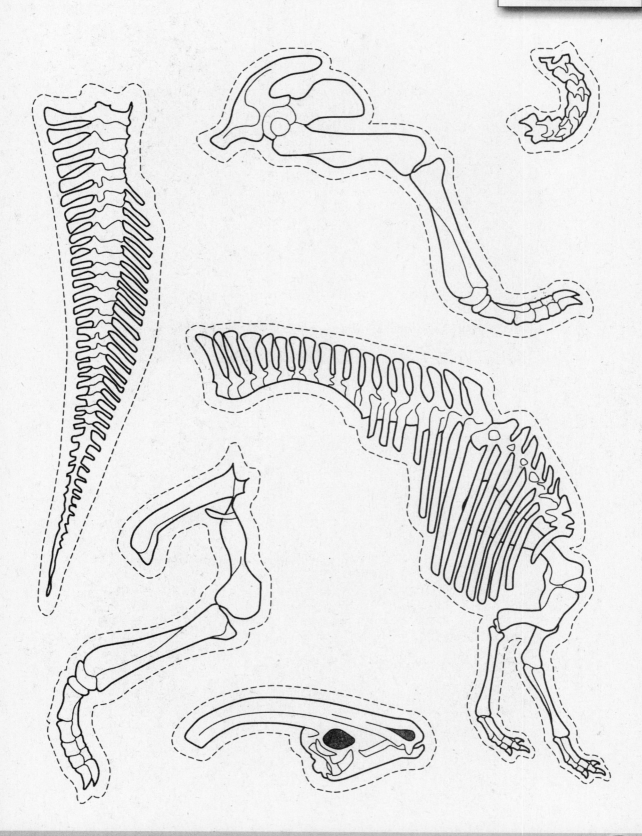

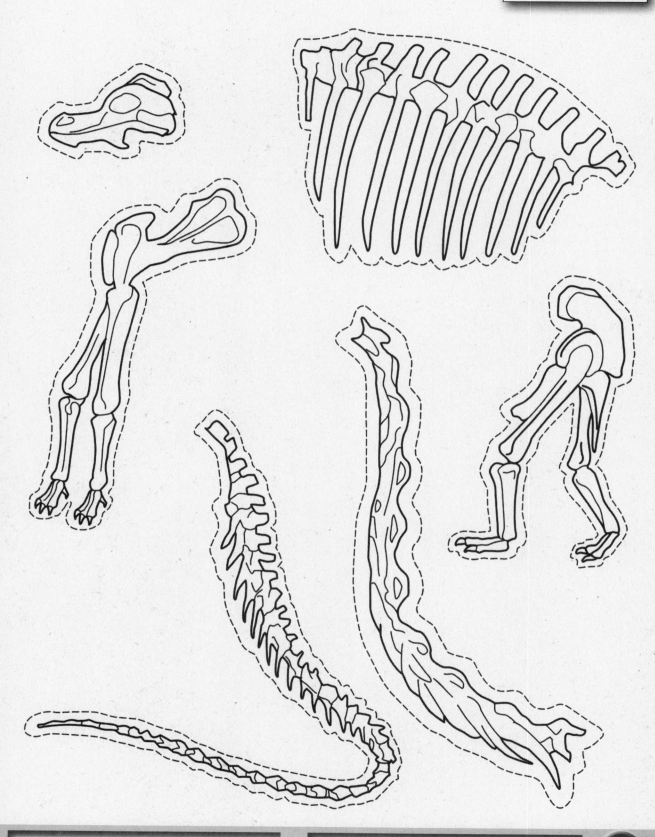

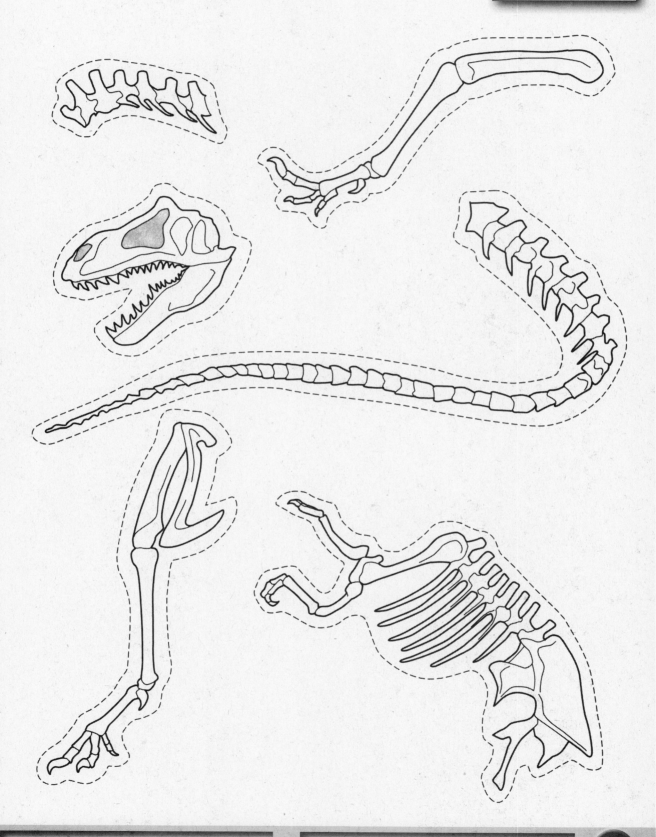

JUNE

Earth

SEPTEMBER

DECEMBER

Puller Pal

Instruction

1. Cut cardboard the same size as the scale on the right.

2. Cut out the scale. Tape it to the cardboard.

3. Use a paper clip to attach a thin rubber band at the top of the scale.

4. Place masking tape on the paper clip to hold the paper clip in place. The end of the rubber band should line up with zero on the scale.

5. Tie a 12-inch string to the end of the rubber band.

— 0
— 1
— 2
— 3
— 4
— 5
— 6
— 7